Technical Studies

INTC 2380

Cooperative Education:
Instrumentation Technology/Technician
Internship Booklet
2020

ISBN-13: 9781686802737

INTC 2380 – Cooperative Education: Instrumentation Technology/Technician

Credits: 3

Career-related activities encountered in the student's area specialization offered through an individualized agreement among the college, employer, and student. Under the supervision of the college and the employer, the student combines classroom learning with work experience. Includes a lecture component.

Lecture Hours 1
Lab Hours 0
External Hours 19

Pre-requisite: INTC 1441 Principles of Automatic Control
READ 300/REBR 300 or equivalent

Instructor: ______________________________

Office Location: ______________________________

Office Hours: ______________________________

Instructor Contact: ______________________________

Meeting time: ______________________________

Division Chair: ______________________________

Division Contact: ______________________________

Dean: ______________________________

Dean Contact: ______________________________

Training Agreement

Program Information:

Semester/Year: ______________________________

Student Name: ______________________________

Company Name: ______________________________

Worksite Address: ______________________________

Worksite City/Zip: ______________________________

Student Position Title: ______________________________

Evaluating Supervisor: ______________________________

Supervisor Contact: ______________________________

Instructor/Coordinator: ______________________________

The student, the evaluating supervisor, and the instructor/coordinator will all cooperate to determine the learning objectives to meet the Student Learning Outcomes (SLOs) listed in the course syllabus. Achievement of the objectives will be part of the basis for the credit and grade, which will be earned for the work experience. The instructor/coordinator will visit with the student on a weekly basis, at the time and location of the instructor/coordinator's choosing, preferably determined in coordination with the schedule of the participating student. The instructor/coordinator will visit with the student and the evaluating supervisor at the job site a minimum of twice per internship semester. The supervisor assists with the evaluation of the student's performance, but the instructor/coordinator will determine the student's final grade.

This document is not a legal contract, and may be terminated at the discretion of the employer or the instructor/coordinator.

Signatures:

Student Signature: ______________________________

Evaluating Supervisor: ______________________________

Instructor/Coordinator: ______________________________

Career & Transfer Services
Student Center 104 & 106

Lena Yepez
lyepez@lee.edu
PO Box 818, Baytown, TX 77520
832-556-4009

Career & Transfer Services provides resources and programs to help students explore career options and develop effective job search skills, enabling them to professionally represent themselves to prospective employers and ultimately attain their employment goals.

The following services are available to Lee College students and alumni:

Career Exploration

Students can meet with the Career & Transfer Services specialist to discuss concerns related to career options, employment trends, gaining experience, preparing for a job search, or making the transition from college to career.

Resumes, Cover Letters, Job Search Correspondence

Career & Transfer Services can help students create resumes, cover letters and thank you notes to prospective employers.

Practice Interviews

Simulated job interviews give a student the opportunity to sharpen his or her interviewing skills. Post interview critiques help students gauge their performance and learn more about effective interview strategies.

Career-Related Workshops and Presentations

Career & Transfer Services hosts a variety of workshops and presentations designed to help students improve their search skills. Topics include developing resumes, writing cover letters, creating a portfolio, preparing for an interview, and others.

Job Fairs

Job fairs provide students and alumni with the opportunity to network, gather information, determine career options, and promote themselves in a less formal environment.

Workplace Information

Company Name: ______________________________

Company Headquarters: ______________________________

Year Company Founded: ______________________________

Primary Products/Services:

____

____

____

Local Worksite Manager: ______________________________

Year Worksite Built: ______________________________

No. of Employees at worksite: ______________________________

Employment Start Date: ______________________________

Internship Start Date: ______________________________

Internship End Date: ______________________________

Typical Work Schedule:

	Sunday	Monday	Tuesday	Wednesday	Thursday	Friday	Saturday
Start Time							
End Time							

Week 1

Work Order # ____________________

Work Summary: ____________________

Duration of Work: ______________________________

Lessons Learned: ______________________________

Work Categories (circle all that apply):

JSA/TSTI/JHA	Hot Work Permit	Excavation Permit	Confined Space Entry
LOTO	Instrumentation	Electrical	Inst. Installation
Elec. Installation	Inst. Removal	Elec. Removal	Inst. Maintenance
Elec. Maintenance	Work Area Cleanup	Inventory Mgt.	Inst. Instruction
Elec. Instruction	Safety Instruction	P&ID	PFD
Wiring Diagram	Motors	DC < 28VDC	DC > 28VDC
AC < 240VAC	AC > 240VAC	Vehicle	Multimeter
Megohmeter	Hand Tools	Power Tools	Tube Bending
Conduit Bending	Conduit Sealing	Communicator	Transmitter calibration
Level Inst.	Pressure Inst.	Temperature Inst.	Flow Inst.
Production	Service	Sales	DCS
HRC1	HRC2	HRC3	HRC4
Parts Ordered	Overhead Lift	Analyzers	PLC

Supervisor Signature: ______________________________

Date: ______________________________

Week 1

Work Order # ____

Work Summary: ____

Duration of Work: ______________________________

Lessons Learned: ______________________________

Work Categories (circle all that apply):

JSA/TSTI/JHA	Hot Work Permit	Excavation Permit	Confined Space Entry
LOTO	Instrumentation	Electrical	Inst. Installation
Elec. Installation	Inst. Removal	Elec. Removal	Inst. Maintenance
Elec. Maintenance	Work Area Cleanup	Inventory Mgt.	Inst. Instruction
Elec. Instruction	Safety Instruction	P&ID	PFD
Wiring Diagram	Motors	DC < 28VDC	DC > 28VDC
AC < 240VAC	AC > 240VAC	Vehicle	Multimeter
Megohmeter	Hand Tools	Power Tools	Tube Bending
Conduit Bending	Conduit Sealing	Communicator	Transmitter calibration
Level Inst.	Pressure Inst.	Temperature Inst.	Flow Inst.
Production	Service	Sales	DCS
HRC1	HRC2	HRC3	HRC4
Parts Ordered	Overhead Lift	Analyzers	PLC

Supervisor Signature: ______________________________

Date: ______________________________

Week 1

Work Order # ______________________________

Work Summary: ______________________________

Duration of Work: ____________________

Lessons Learned: ____________________

Work Categories (circle all that apply):

JSA/TSTI/JHA	Hot Work Permit	Excavation Permit	Confined Space Entry
LOTO	Instrumentation	Electrical	Inst. Installation
Elec. Installation	Inst. Removal	Elec. Removal	Inst. Maintenance
Elec. Maintenance	Work Area Cleanup	Inventory Mgt.	Inst. Instruction
Elec. Instruction	Safety Instruction	P&ID	PFD
Wiring Diagram	Motors	DC < 28VDC	DC > 28VDC
AC < 240VAC	AC > 240VAC	Vehicle	Multimeter
Megohmeter	Hand Tools	Power Tools	Tube Bending
Conduit Bending	Conduit Sealing	Communicator	Transmitter calibration
Level Inst.	Pressure Inst.	Temperature Inst.	Flow Inst.
Production	Service	Sales	DCS
HRC1	HRC2	HRC3	HRC4
Parts Ordered	Overhead Lift	Analyzers	PLC

Supervisor Signature: ______________________________

Date: ______________________________

Week 1

Work Order # ______________________________

Work Summary: ______________________________

Duration of Work: ______________________________

Lessons Learned: ______________________________

Work Categories (circle all that apply):

JSA/TSTI/JHA	Hot Work Permit	Excavation Permit	Confined Space Entry
LOTO	Instrumentation	Electrical	Inst. Installation
Elec. Installation	Inst. Removal	Elec. Removal	Inst. Maintenance
Elec. Maintenance	Work Area Cleanup	Inventory Mgt.	Inst. Instruction
Elec. Instruction	Safety Instruction	P&ID	PFD
Wiring Diagram	Motors	DC < 28VDC	DC > 28VDC
AC < 240VAC	AC > 240VAC	Vehicle	Multimeter
Megohmeter	Hand Tools	Power Tools	Tube Bending
Conduit Bending	Conduit Sealing	Communicator	Transmitter calibration
Level Inst.	Pressure Inst.	Temperature Inst.	Flow Inst.
Production	Service	Sales	DCS
HRC1	HRC2	HRC3	HRC4
Parts Ordered	Overhead Lift	Analyzers	PLC

Supervisor Signature: ______________________________

Date: ______________________________

Week 2

Work Order # ____________________

Work Summary: ____________________

Duration of Work: ______________________________

Lessons Learned: ______________________________

Work Categories (circle all that apply):

JSA/TSTI/JHA	Hot Work Permit	Excavation Permit	Confined Space Entry
LOTO	Instrumentation	Electrical	Inst. Installation
Elec. Installation	Inst. Removal	Elec. Removal	Inst. Maintenance
Elec. Maintenance	Work Area Cleanup	Inventory Mgt.	Inst. Instruction
Elec. Instruction	Safety Instruction	P&ID	PFD
Wiring Diagram	Motors	DC < 28VDC	DC > 28VDC
AC < 240VAC	AC > 240VAC	Vehicle	Multimeter
Megohmeter	Hand Tools	Power Tools	Tube Bending
Conduit Bending	Conduit Sealing	Communicator	Transmitter calibration
Level Inst.	Pressure Inst.	Temperature Inst.	Flow Inst.
Production	Service	Sales	DCS
HRC1	HRC2	HRC3	HRC4
Parts Ordered	Overhead Lift	Analyzers	PLC

Supervisor Signature: ______________________________

Date: ______________________________

Week 2

Work Order # ____________________

Work Summary: ____________________

Duration of Work: ______________________________

Lessons Learned: ______________________________

Work Categories (circle all that apply):

JSA/TSTI/JHA	Hot Work Permit	Excavation Permit	Confined Space Entry
LOTO	Instrumentation	Electrical	Inst. Installation
Elec. Installation	Inst. Removal	Elec. Removal	Inst. Maintenance
Elec. Maintenance	Work Area Cleanup	Inventory Mgt.	Inst. Instruction
Elec. Instruction	Safety Instruction	P&ID	PFD
Wiring Diagram	Motors	DC < 28VDC	DC > 28VDC
AC < 240VAC	AC > 240VAC	Vehicle	Multimeter
Megohmeter	Hand Tools	Power Tools	Tube Bending
Conduit Bending	Conduit Sealing	Communicator	Transmitter calibration
Level Inst.	Pressure Inst.	Temperature Inst.	Flow Inst.
Production	Service	Sales	DCS
HRC1	HRC2	HRC3	HRC4
Parts Ordered	Overhead Lift	Analyzers	PLC

Supervisor Signature: ______________________________

Date: ______________________________

Week 2

Work Order # 910255124-0010

Work Summary: V5307 Replace Tubing

PDI-249 Tubing + Transmitter needs to be replaced has Multiple leaks on Tubing and fittings.

Duration of Work: 5 hrs.

Lessons Learned: 1. Always verify line break

2. Test before you disconnect

3. Always fill out JSA prior to start work.

4. Always have your parts ordered.

5. Use the right tools for the job.

Work Categories (circle all that apply):

JSA/TSTI/JHA	Hot Work Permit	Excavation Permit	Confined Space Entry
LOTO	Instrumentation	Electrical	Inst. Installation
Elec. Installation	Inst. Removal	Elec. Removal	Inst. Maintenance
Elec. Maintenance	Work Area Cleanup	Inventory Mgt.	Inst. Instruction
Elec. Instruction	Safety Instruction	P&ID	PFD
Wiring Diagram	Motors	DC < 28VDC	DC > 28VDC
AC < 240VAC	AC > 240VAC	Vehicle	Multimeter
Megohmeter	Hand Tools	Power Tools	Tube Bending
Conduit Bending	Conduit Sealing	Communicator	Transmitter calibration
Level Inst.	Pressure Inst.	Temperature Inst.	Flow Inst.
Production	Service	Sales	DCS
HRC1	HRC2	HRC3	HRC4
Parts Ordered	Overhead Lift	Analyzers	PLC

Supervisor Signature: ______________________

Date: 02/10/20

Week 2

Work Order #

Work Summary:

Duration of Work: ____________________

Lessons Learned: ____________________

Work Categories (circle all that apply):

JSA/TSTI/JHA	Hot Work Permit	Excavation Permit	Confined Space Entry
LOTO	Instrumentation	Electrical	Inst. Installation
Elec. Installation	Inst. Removal	Elec. Removal	Inst. Maintenance
Elec. Maintenance	Work Area Cleanup	Inventory Mgt.	Inst. Instruction
Elec. Instruction	Safety Instruction	P&ID	PFD
Wiring Diagram	Motors	DC < 28VDC	DC > 28VDC
AC < 240VAC	AC > 240VAC	Vehicle	Multimeter
Megohmeter	Hand Tools	Power Tools	Tube Bending
Conduit Bending	Conduit Sealing	Communicator	Transmitter calibration
Level Inst.	Pressure Inst.	Temperature Inst.	Flow Inst.
Production	Service	Sales	DCS
HRC1	HRC2	HRC3	HRC4
Parts Ordered	Overhead Lift	Analyzers	PLC

Supervisor Signature: ____

Date: ____

Week 3

Work Order # 910250290 -0010

Work Summary: 500 Train Troubleshoot Alley Lighting

Alleyway 500 has several lights not working properly. Need I/E to troubleshoot and repair as needed.

Duration of Work: 5 hrs.

Lessons Learned:

1. Always fill JSA in the field
2. LOTO-OUT circuits to work-on
3. Remove old Lights
4. follow safety Instructions
5. Always verify with wiring Diagrams

Work Categories (circle all that apply):

JSA/TSTI/JHA	Hot Work Permit	Excavation Permit	Confined Space Entry
LOTO	Instrumentation	Electrical	Inst. Installation
Elec. Installation	Inst. Removal	Elec. Removal	Inst. Maintenance
Elec. Maintenance	Work Area Cleanup	Inventory Mgt.	Inst. Instruction
Elec. Instruction	Safety Instruction	P&ID	PFD
Wiring Diagram	Motors	DC < 28VDC	DC > 28VDC
AC < 240VAC	AC > 240VAC	Vehicle	Multimeter
Megohmeter	Hand Tools	Power Tools	Tube Bending
Conduit Bending	Conduit Sealing	Communicator	Transmitter calibration
Level Inst.	Pressure Inst.	Temperature Inst.	Flow Inst.
Production	Service	Sales	DCS
HRC1	HRC2	HRC3	HRC4
Parts Ordered	Overhead Lift	Analyzers	PLC

Supervisor Signature: ______________________

Date: 02/17/20

Week 3

Work Order # ______________________________

Work Summary: ______________________________

Duration of Work: ____

Lessons Learned: ____

Work Categories (circle all that apply):

JSA/TSTI/JHA	Hot Work Permit	Excavation Permit	Confined Space Entry
LOTO	Instrumentation	Electrical	Inst. Installation
Elec. Installation	Inst. Removal	Elec. Removal	Inst. Maintenance
Elec. Maintenance	Work Area Cleanup	Inventory Mgt.	Inst. Instruction
Elec. Instruction	Safety Instruction	P&ID	PFD
Wiring Diagram	Motors	DC < 28VDC	DC > 28VDC
AC < 240VAC	AC > 240VAC	Vehicle	Multimeter
Megohmeter	Hand Tools	Power Tools	Tube Bending
Conduit Bending	Conduit Sealing	Communicator	Transmitter calibration
Level Inst.	Pressure Inst.	Temperature Inst.	Flow Inst.
Production	Service	Sales	DCS
HRC1	HRC2	HRC3	HRC4
Parts Ordered	Overhead Lift	Analyzers	PLC

Supervisor Signature: ______________________________

Date: ______________________________

Week 3

Work Order # ______________________________

Work Summary: ______________________________

Duration of Work: ______________________________

Lessons Learned: ______________________________

Work Categories (circle all that apply):

JSA/TSTI/JHA	Hot Work Permit	Excavation Permit	Confined Space Entry
LOTO	Instrumentation	Electrical	Inst. Installation
Elec. Installation	Inst. Removal	Elec. Removal	Inst. Maintenance
Elec. Maintenance	Work Area Cleanup	Inventory Mgt.	Inst. Instruction
Elec. Instruction	Safety Instruction	P&ID	PFD
Wiring Diagram	Motors	DC < 28VDC	DC > 28VDC
AC < 240VAC	AC > 240VAC	Vehicle	Multimeter
Megohmeter	Hand Tools	Power Tools	Tube Bending
Conduit Bending	Conduit Sealing	Communicator	Transmitter calibration
Level Inst.	Pressure Inst.	Temperature Inst.	Flow Inst.
Production	Service	Sales	DCS
HRC1	HRC2	HRC3	HRC4
Parts Ordered	Overhead Lift	Analyzers	PLC

Supervisor Signature: ______________________________

Date: ______________________________

Week 3

Work Order # ______________________________

Work Summary: ______________________________

Duration of Work: ____________________

Lessons Learned: ____________________

Work Categories (circle all that apply):

JSA/TSTI/JHA	Hot Work Permit	Excavation Permit	Confined Space Entry
LOTO	Instrumentation	Electrical	Inst. Installation
Elec. Installation	Inst. Removal	Elec. Removal	Inst. Maintenance
Elec. Maintenance	Work Area Cleanup	Inventory Mgt.	Inst. Instruction
Elec. Instruction	Safety Instruction	P&ID	PFD
Wiring Diagram	Motors	DC < 28VDC	DC > 28VDC
AC < 240VAC	AC > 240VAC	Vehicle	Multimeter
Megohmeter	Hand Tools	Power Tools	Tube Bending
Conduit Bending	Conduit Sealing	Communicator	Transmitter calibration
Level Inst.	Pressure Inst.	Temperature Inst.	Flow Inst.
Production	Service	Sales	DCS
HRC1	HRC2	HRC3	HRC4
Parts Ordered	Overhead Lift	Analyzers	PLC

Supervisor Signature: ______________________________

Date: ______________________________

Week 4

Work Order # 910246254 - 10

Work Summary: K12 - Replace lights fire station

Troubleshoot and Replace lights inside and Outside firestation #12 using JLG and/or Scissor Lift.

Duration of Work: 36 hours.

Lessons Learned:

1. Always Verify LOTO
2. Test before you work
3. Use Multimeter to verify Voltage
4. Always use the proper tools for the job
5. Use Wiring Diagrams.

Work Categories (circle all that apply):

JSA/TSTI/JHA	Hot Work Permit	Excavation Permit	Confined Space Entry
LOTO	Instrumentation	Electrical	Inst. Installation
Elec. Installation	Inst. Removal	Elec. Removal	Inst. Maintenance
Elec. Maintenance	Work Area Cleanup	Inventory Mgt.	Inst. Instruction
Elec. Instruction	Safety Instruction	P&ID	PFD
Wiring Diagram	Motors	DC < 28VDC	DC > 28VDC
AC < 240VAC	AC > 240VAC	Vehicle	Multimeter
Megohmeter	Hand Tools	Power Tools	Tube Bending
Conduit Bending	Conduit Sealing	Communicator	Transmitter calibration
Level Inst.	Pressure Inst.	Temperature Inst.	Flow Inst.
Production	Service	Sales	DCS
HRC1	HRC2	HRC3	HRC4
Parts Ordered	Overhead Lift	Analyzers	PLC

Supervisor Signature: ______________________

Date: 02/24/20

Week 4

Work Order # ______________________________

Work Summary: ______________________________

Duration of Work:

Lessons Learned:

Work Categories (circle all that apply):

JSA/TSTI/JHA	Hot Work Permit	Excavation Permit	Confined Space Entry
LOTO	Instrumentation	Electrical	Inst. Installation
Elec. Installation	Inst. Removal	Elec. Removal	Inst. Maintenance
Elec. Maintenance	Work Area Cleanup	Inventory Mgt.	Inst. Instruction
Elec. Instruction	Safety Instruction	P&ID	PFD
Wiring Diagram	Motors	DC < 28VDC	DC > 28VDC
AC < 240VAC	AC > 240VAC	Vehicle	Multimeter
Megohmeter	Hand Tools	Power Tools	Tube Bending
Conduit Bending	Conduit Sealing	Communicator	Transmitter calibration
Level Inst.	Pressure Inst.	Temperature Inst.	Flow Inst.
Production	Service	Sales	DCS
HRC1	HRC2	HRC3	HRC4
Parts Ordered	Overhead Lift	Analyzers	PLC

Supervisor Signature: ______________________________

Date: ______________________________

Week 4

Work Order #

Work Summary:

Duration of Work: ______________________________

Lessons Learned: ______________________________

Work Categories (circle all that apply):

JSA/TSTI/JHA	Hot Work Permit	Excavation Permit	Confined Space Entry
LOTO	Instrumentation	Electrical	Inst. Installation
Elec. Installation	Inst. Removal	Elec. Removal	Inst. Maintenance
Elec. Maintenance	Work Area Cleanup	Inventory Mgt.	Inst. Instruction
Elec. Instruction	Safety Instruction	P&ID	PFD
Wiring Diagram	Motors	DC < 28VDC	DC > 28VDC
AC < 240VAC	AC > 240VAC	Vehicle	Multimeter
Megohmeter	Hand Tools	Power Tools	Tube Bending
Conduit Bending	Conduit Sealing	Communicator	Transmitter calibration
Level Inst.	Pressure Inst.	Temperature Inst.	Flow Inst.
Production	Service	Sales	DCS
HRC1	HRC2	HRC3	HRC4
Parts Ordered	Overhead Lift	Analyzers	PLC

Supervisor Signature: ______________________________

Date: ______________________________

Week 4

Work Order # ______________________________

Work Summary: ______________________________

Duration of Work: ______________________________

Lessons Learned: ______________________________

Work Categories (circle all that apply):

JSA/TSTI/JHA	Hot Work Permit	Excavation Permit	Confined Space Entry
LOTO	Instrumentation	Electrical	Inst. Installation
Elec. Installation	Inst. Removal	Elec. Removal	Inst. Maintenance
Elec. Maintenance	Work Area Cleanup	Inventory Mgt.	Inst. Instruction
Elec. Instruction	Safety Instruction	P&ID	PFD
Wiring Diagram	Motors	DC < 28VDC	DC > 28VDC
AC < 240VAC	AC > 240VAC	Vehicle	Multimeter
Megohmeter	Hand Tools	Power Tools	Tube Bending
Conduit Bending	Conduit Sealing	Communicator	Transmitter calibration
Level Inst.	Pressure Inst.	Temperature Inst.	Flow Inst.
Production	Service	Sales	DCS
HRC1	HRC2	HRC3	HRC4
Parts Ordered	Overhead Lift	Analyzers	PLC

Supervisor Signature: ______________________________

Date: ______________________________

4 Week Evaluating Supervisor Report

Evaluating Supervisor: Dwight Ottelle

Date: 03/02/20

1. Follows safe work practices, follows rules, does not take short cuts, and understands company safety policies.

   a. Excellent (circled)
   b. Very Good
   c. Good
   d. Satisfactory
   e. Unsatisfactory

2. Ability to adapt to and work effectively in a variety of situations and with various individuals and groups. Adapts one's approach as the requirements of the situation change.

   a. Excellent (circled)
   b. Very Good
   c. Good
   d. Satisfactory
   e. Unsatisfactory

3. Self-motivated to learn and share ideas. Driven by curiosity and a desire to know more about the unit process, equipment, people and Health, Safety, and Environmental (HSE) initiatives.

   a. Excellent (circled)
   b. Very Good
   c. Good
   d. Satisfactory
   e. Unsatisfactory

4. Good attendance. Always at work and on time unless sick or an excusable emergency.

   a. Excellent (circled)
   b. Very Good
   c. Good
   d. Satisfactory
   e. Unsatisfactory

5. Gets along with others, wants to be a part of the team as opposed to working sepa4rately. Supports the team concept by information sharing, listening, and contributing ideas.

    a. Excellent
    b. Very Good
    c. Good
    d. Satisfactory
    e. Unsatisfactory

6. Respects the meaning of privileged information. Follows established company procedures.

    a. Excellent
    b. Very Good
    c. Good
    d. Satisfactory
    e. Unsatisfactory

7. Maintains/Prepares satisfactory records. Recognizes tasks that are beyond student capacity. Identifies and attempts to solve discrepancies in systems, results, or information.

    a. Excellent
    b. Very Good
    c. Good
    d. Satisfactory
    e. Unsatisfactory

8. Starts activities in a timely manner. Willingly stays to complete or correct work. Organizes workload.

    a. Excellent
    b. Very Good
    c. Good
    d. Satisfactory
    e. Unsatisfactory

9. Routine tasks are competed within acceptable time. Transfers knowledge of principles and procedures to new techniques. Approaches assignments with confidence.

    a. Excellent
    b. Very Good
    c. Good
    d. Satisfactory
    e. Unsatisfactory

Week 5

Work Order # 97073306-0060

Work Summary: MOC 93928 10LT-812 Rmv Lvl/Mod. Conduit

Remove existing dp XTMR and Tubing from 10LT-812 and modify conduit for new capillary Transmitter.

Duration of Work: 18 hrs.

Lessons Learned:

1. Always Disconnect and coil and tape signal wire to Rosemount level Transmitter.
2. Always wear a harness
3. Fill-out JSA in the field
4. Communicate Instrument removal with the board operator.

Work Categories (circle all that apply):

JSA/TSTI/JHA	Hot Work Permit	Excavation Permit	Confined Space Entry
LOTO	Instrumentation	Electrical	Inst. Installation
Elec. Installation	Inst. Removal	Elec. Removal	Inst. Maintenance
Elec. Maintenance	Work Area Cleanup	Inventory Mgt.	Inst. Instruction
Elec. Instruction	Safety Instruction	P&ID	PFD
Wiring Diagram	Motors	DC < 28VDC	DC > 28VDC
AC < 240VAC	AC > 240VAC	Vehicle	Multimeter
Megohmeter	Hand Tools	Power Tools	Tube Bending
Conduit Bending	Conduit Sealing	Communicator	Transmitter calibration
Level Inst.	Pressure Inst.	Temperature Inst.	Flow Inst.
Production	Service	Sales	DCS
HRC1	HRC2	HRC3	HRC4
Parts Ordered	Overhead Lift	Analyzers	PLC

Supervisor Signature: ______________________

Date: 03/16/20

Week 5

Work Order # ____

Work Summary: ____

Duration of Work: ____

Lessons Learned: ____

Work Categories (circle all that apply):

JSA/TSTI/JHA	Hot Work Permit	Excavation Permit	Confined Space Entry
LOTO	Instrumentation	Electrical	Inst. Installation
Elec. Installation	Inst. Removal	Elec. Removal	Inst. Maintenance
Elec. Maintenance	Work Area Cleanup	Inventory Mgt.	Inst. Instruction
Elec. Instruction	Safety Instruction	P&ID	PFD
Wiring Diagram	Motors	DC < 28VDC	DC > 28VDC
AC < 240VAC	AC > 240VAC	Vehicle	Multimeter
Megohmeter	Hand Tools	Power Tools	Tube Bending
Conduit Bending	Conduit Sealing	Communicator	Transmitter calibration
Level Inst.	Pressure Inst.	Temperature Inst.	Flow Inst.
Production	Service	Sales	DCS
HRC1	HRC2	HRC3	HRC4
Parts Ordered	Overhead Lift	Analyzers	PLC

Supervisor Signature: ______________________________

Date: ______________________________

Week 5

Work Order # ______________________________

Work Summary: ______________________________

Duration of Work: ____

Lessons Learned: ____

Work Categories (circle all that apply):

JSA/TSTI/JHA	Hot Work Permit	Excavation Permit	Confined Space Entry
LOTO	Instrumentation	Electrical	Inst. Installation
Elec. Installation	Inst. Removal	Elec. Removal	Inst. Maintenance
Elec. Maintenance	Work Area Cleanup	Inventory Mgt.	Inst. Instruction
Elec. Instruction	Safety Instruction	P&ID	PFD
Wiring Diagram	Motors	DC < 28VDC	DC > 28VDC
AC < 240VAC	AC > 240VAC	Vehicle	Multimeter
Megohmeter	Hand Tools	Power Tools	Tube Bending
Conduit Bending	Conduit Sealing	Communicator	Transmitter calibration
Level Inst.	Pressure Inst.	Temperature Inst.	Flow Inst.
Production	Service	Sales	DCS
HRC1	HRC2	HRC3	HRC4
Parts Ordered	Overhead Lift	Analyzers	PLC

Supervisor Signature: ____

Date: ____

Week 5

Work Order #

Work Summary:

Duration of Work: ____________________

Lessons Learned: ____________________

Work Categories (circle all that apply):

JSA/TSTI/JHA	Hot Work Permit	Excavation Permit	Confined Space Entry
LOTO	Instrumentation	Electrical	Inst. Installation
Elec. Installation	Inst. Removal	Elec. Removal	Inst. Maintenance
Elec. Maintenance	Work Area Cleanup	Inventory Mgt.	Inst. Instruction
Elec. Instruction	Safety Instruction	P&ID	PFD
Wiring Diagram	Motors	DC < 28VDC	DC > 28VDC
AC < 240VAC	AC > 240VAC	Vehicle	Multimeter
Megohmeter	Hand Tools	Power Tools	Tube Bending
Conduit Bending	Conduit Sealing	Communicator	Transmitter calibration
Level Inst.	Pressure Inst.	Temperature Inst.	Flow Inst.
Production	Service	Sales	DCS
HRC1	HRC2	HRC3	HRC4
Parts Ordered	Overhead Lift	Analyzers	PLC

Supervisor Signature: ____

Date: ____

Week 6

Work Order # ______________________

Work Summary: ______________________

Duration of Work: ______________________________

Lessons Learned: ______________________________

Work Categories (circle all that apply):

JSA/TSTI/JHA	Hot Work Permit	Excavation Permit	Confined Space Entry
LOTO	Instrumentation	Electrical	Inst. Installation
Elec. Installation	Inst. Removal	Elec. Removal	Inst. Maintenance
Elec. Maintenance	Work Area Cleanup	Inventory Mgt.	Inst. Instruction
Elec. Instruction	Safety Instruction	P&ID	PFD
Wiring Diagram	Motors	DC < 28VDC	DC > 28VDC
AC < 240VAC	AC > 240VAC	Vehicle	Multimeter
Megohmeter	Hand Tools	Power Tools	Tube Bending
Conduit Bending	Conduit Sealing	Communicator	Transmitter calibration
Level Inst.	Pressure Inst.	Temperature Inst.	Flow Inst.
Production	Service	Sales	DCS
HRC1	HRC2	HRC3	HRC4
Parts Ordered	Overhead Lift	Analyzers	PLC

Supervisor Signature: ______________________________

Date: ______________________________

Week 6

Work Order # ______________________________

Work Summary: ______________________________

Duration of Work: ______________________________

Lessons Learned: ______________________________

Work Categories (circle all that apply):

JSA/TSTI/JHA	Hot Work Permit	Excavation Permit	Confined Space Entry
LOTO	Instrumentation	Electrical	Inst. Installation
Elec. Installation	Inst. Removal	Elec. Removal	Inst. Maintenance
Elec. Maintenance	Work Area Cleanup	Inventory Mgt.	Inst. Instruction
Elec. Instruction	Safety Instruction	P&ID	PFD
Wiring Diagram	Motors	DC < 28VDC	DC > 28VDC
AC < 240VAC	AC > 240VAC	Vehicle	Multimeter
Megohmeter	Hand Tools	Power Tools	Tube Bending
Conduit Bending	Conduit Sealing	Communicator	Transmitter calibration
Level Inst.	Pressure Inst.	Temperature Inst.	Flow Inst.
Production	Service	Sales	DCS
HRC1	HRC2	HRC3	HRC4
Parts Ordered	Overhead Lift	Analyzers	PLC

Supervisor Signature: ______________________________

Date: ______________________________

Week 6

Work Order #

Work Summary:

Duration of Work: ______________________________

Lessons Learned: ______________________________

Work Categories (circle all that apply):

JSA/TSTI/JHA	Hot Work Permit	Excavation Permit	Confined Space Entry
LOTO	Instrumentation	Electrical	Inst. Installation
Elec. Installation	Inst. Removal	Elec. Removal	Inst. Maintenance
Elec. Maintenance	Work Area Cleanup	Inventory Mgt.	Inst. Instruction
Elec. Instruction	Safety Instruction	P&ID	PFD
Wiring Diagram	Motors	DC < 28VDC	DC > 28VDC
AC < 240VAC	AC > 240VAC	Vehicle	Multimeter
Megohmeter	Hand Tools	Power Tools	Tube Bending
Conduit Bending	Conduit Sealing	Communicator	Transmitter calibration
Level Inst.	Pressure Inst.	Temperature Inst.	Flow Inst.
Production	Service	Sales	DCS
HRC1	HRC2	HRC3	HRC4
Parts Ordered	Overhead Lift	Analyzers	PLC

Supervisor Signature: ______________________________

Date: ______________________________

Week 6

Work Order #

Work Summary:

Duration of Work: ______________________________

Lessons Learned: ______________________________

Work Categories (circle all that apply):

JSA/TSTI/JHA	Hot Work Permit	Excavation Permit	Confined Space Entry
LOTO	Instrumentation	Electrical	Inst. Installation
Elec. Installation	Inst. Removal	Elec. Removal	Inst. Maintenance
Elec. Maintenance	Work Area Cleanup	Inventory Mgt.	Inst. Instruction
Elec. Instruction	Safety Instruction	P&ID	PFD
Wiring Diagram	Motors	DC < 28VDC	DC > 28VDC
AC < 240VAC	AC > 240VAC	Vehicle	Multimeter
Megohmeter	Hand Tools	Power Tools	Tube Bending
Conduit Bending	Conduit Sealing	Communicator	Transmitter calibration
Level Inst.	Pressure Inst.	Temperature Inst.	Flow Inst.
Production	Service	Sales	DCS
HRC1	HRC2	HRC3	HRC4
Parts Ordered	Overhead Lift	Analyzers	PLC

Supervisor Signature: ______________________________

Date: ______________________________

Week 7

Work Order # ____________________

Work Summary: ____________________

Duration of Work: ____________________

Lessons Learned: ____________________

Work Categories (circle all that apply):

JSA/TSTI/JHA	Hot Work Permit	Excavation Permit	Confined Space Entry
LOTO	Instrumentation	Electrical	Inst. Installation
Elec. Installation	Inst. Removal	Elec. Removal	Inst. Maintenance
Elec. Maintenance	Work Area Cleanup	Inventory Mgt.	Inst. Instruction
Elec. Instruction	Safety Instruction	P&ID	PFD
Wiring Diagram	Motors	DC < 28VDC	DC > 28VDC
AC < 240VAC	AC > 240VAC	Vehicle	Multimeter
Megohmeter	Hand Tools	Power Tools	Tube Bending
Conduit Bending	Conduit Sealing	Communicator	Transmitter calibration
Level Inst.	Pressure Inst.	Temperature Inst.	Flow Inst.
Production	Service	Sales	DCS
HRC1	HRC2	HRC3	HRC4
Parts Ordered	Overhead Lift	Analyzers	PLC

Supervisor Signature: ______________________________

Date: ______________________________

Week 7

Work Order # ____________________

Work Summary: ____________________

Duration of Work: ____________________

Lessons Learned: ____________________

Work Categories (circle all that apply):

JSA/TSTI/JHA	Hot Work Permit	Excavation Permit	Confined Space Entry
LOTO	Instrumentation	Electrical	Inst. Installation
Elec. Installation	Inst. Removal	Elec. Removal	Inst. Maintenance
Elec. Maintenance	Work Area Cleanup	Inventory Mgt.	Inst. Instruction
Elec. Instruction	Safety Instruction	P&ID	PFD
Wiring Diagram	Motors	DC < 28VDC	DC > 28VDC
AC < 240VAC	AC > 240VAC	Vehicle	Multimeter
Megohmeter	Hand Tools	Power Tools	Tube Bending
Conduit Bending	Conduit Sealing	Communicator	Transmitter calibration
Level Inst.	Pressure Inst.	Temperature Inst.	Flow Inst.
Production	Service	Sales	DCS
HRC1	HRC2	HRC3	HRC4
Parts Ordered	Overhead Lift	Analyzers	PLC

Supervisor Signature: ______________________________

Date: ______________________________

Week 7

Work Order # ______________________________

Work Summary: ______________________________

Duration of Work: ______________________________

Lessons Learned: ______________________________

Work Categories (circle all that apply):

JSA/TSTI/JHA	Hot Work Permit	Excavation Permit	Confined Space Entry
LOTO	Instrumentation	Electrical	Inst. Installation
Elec. Installation	Inst. Removal	Elec. Removal	Inst. Maintenance
Elec. Maintenance	Work Area Cleanup	Inventory Mgt.	Inst. Instruction
Elec. Instruction	Safety Instruction	P&ID	PFD
Wiring Diagram	Motors	DC < 28VDC	DC > 28VDC
AC < 240VAC	AC > 240VAC	Vehicle	Multimeter
Megohmeter	Hand Tools	Power Tools	Tube Bending
Conduit Bending	Conduit Sealing	Communicator	Transmitter calibration
Level Inst.	Pressure Inst.	Temperature Inst.	Flow Inst.
Production	Service	Sales	DCS
HRC1	HRC2	HRC3	HRC4
Parts Ordered	Overhead Lift	Analyzers	PLC

Supervisor Signature: ______________________________

Date: ______________________________

Week 7

Work Order # ____________________

Work Summary: ____________________

Duration of Work: ____________________

Lessons Learned: ____________________

Work Categories (circle all that apply):

JSA/TSTI/JHA	Hot Work Permit	Excavation Permit	Confined Space Entry
LOTO	Instrumentation	Electrical	Inst. Installation
Elec. Installation	Inst. Removal	Elec. Removal	Inst. Maintenance
Elec. Maintenance	Work Area Cleanup	Inventory Mgt.	Inst. Instruction
Elec. Instruction	Safety Instruction	P&ID	PFD
Wiring Diagram	Motors	DC < 28VDC	DC > 28VDC
AC < 240VAC	AC > 240VAC	Vehicle	Multimeter
Megohmeter	Hand Tools	Power Tools	Tube Bending
Conduit Bending	Conduit Sealing	Communicator	Transmitter calibration
Level Inst.	Pressure Inst.	Temperature Inst.	Flow Inst.
Production	Service	Sales	DCS
HRC1	HRC2	HRC3	HRC4
Parts Ordered	Overhead Lift	Analyzers	PLC

Supervisor Signature: ______________________________

Date: ______________________________

Week 8

Work Order # ____________________

Work Summary: ____________________

Duration of Work: ______________________________

Lessons Learned: ______________________________

Work Categories (circle all that apply):

JSA/TSTI/JHA	Hot Work Permit	Excavation Permit	Confined Space Entry
LOTO	Instrumentation	Electrical	Inst. Installation
Elec. Installation	Inst. Removal	Elec. Removal	Inst. Maintenance
Elec. Maintenance	Work Area Cleanup	Inventory Mgt.	Inst. Instruction
Elec. Instruction	Safety Instruction	P&ID	PFD
Wiring Diagram	Motors	DC < 28VDC	DC > 28VDC
AC < 240VAC	AC > 240VAC	Vehicle	Multimeter
Megohmeter	Hand Tools	Power Tools	Tube Bending
Conduit Bending	Conduit Sealing	Communicator	Transmitter calibration
Level Inst.	Pressure Inst.	Temperature Inst.	Flow Inst.
Production	Service	Sales	DCS
HRC1	HRC2	HRC3	HRC4
Parts Ordered	Overhead Lift	Analyzers	PLC

Supervisor Signature: ______________________________

Date: ______________________________

Week 8

Work Order # ______________________________

Work Summary: ______________________________

Duration of Work: ______________________________

Lessons Learned: ______________________________

Work Categories (circle all that apply):

JSA/TSTI/JHA	Hot Work Permit	Excavation Permit	Confined Space Entry
LOTO	Instrumentation	Electrical	Inst. Installation
Elec. Installation	Inst. Removal	Elec. Removal	Inst. Maintenance
Elec. Maintenance	Work Area Cleanup	Inventory Mgt.	Inst. Instruction
Elec. Instruction	Safety Instruction	P&ID	PFD
Wiring Diagram	Motors	DC < 28VDC	DC > 28VDC
AC < 240VAC	AC > 240VAC	Vehicle	Multimeter
Megohmeter	Hand Tools	Power Tools	Tube Bending
Conduit Bending	Conduit Sealing	Communicator	Transmitter calibration
Level Inst.	Pressure Inst.	Temperature Inst.	Flow Inst.
Production	Service	Sales	DCS
HRC1	HRC2	HRC3	HRC4
Parts Ordered	Overhead Lift	Analyzers	PLC

Supervisor Signature: ______________________________

Date: ______________________________

Week 8

Work Order # ______________________________

Work Summary: ______________________________

Duration of Work: __________

Lessons Learned: __________

Work Categories (circle all that apply):

JSA/TSTI/JHA	Hot Work Permit	Excavation Permit	Confined Space Entry
LOTO	Instrumentation	Electrical	Inst. Installation
Elec. Installation	Inst. Removal	Elec. Removal	Inst. Maintenance
Elec. Maintenance	Work Area Cleanup	Inventory Mgt.	Inst. Instruction
Elec. Instruction	Safety Instruction	P&ID	PFD
Wiring Diagram	Motors	DC < 28VDC	DC > 28VDC
AC < 240VAC	AC > 240VAC	Vehicle	Multimeter
Megohmeter	Hand Tools	Power Tools	Tube Bending
Conduit Bending	Conduit Sealing	Communicator	Transmitter calibration
Level Inst.	Pressure Inst.	Temperature Inst.	Flow Inst.
Production	Service	Sales	DCS
HRC1	HRC2	HRC3	HRC4
Parts Ordered	Overhead Lift	Analyzers	PLC

Supervisor Signature: ______________________________

Date: ______________________________

Week 8

Work Order # ______________________________

Work Summary: ______________________________

Duration of Work: ______________________________

Lessons Learned: ______________________________

Work Categories (circle all that apply):

JSA/TSTI/JHA	Hot Work Permit	Excavation Permit	Confined Space Entry
LOTO	Instrumentation	Electrical	Inst. Installation
Elec. Installation	Inst. Removal	Elec. Removal	Inst. Maintenance
Elec. Maintenance	Work Area Cleanup	Inventory Mgt.	Inst. Instruction
Elec. Instruction	Safety Instruction	P&ID	PFD
Wiring Diagram	Motors	DC < 28VDC	DC > 28VDC
AC < 240VAC	AC > 240VAC	Vehicle	Multimeter
Megohmeter	Hand Tools	Power Tools	Tube Bending
Conduit Bending	Conduit Sealing	Communicator	Transmitter calibration
Level Inst.	Pressure Inst.	Temperature Inst.	Flow Inst.
Production	Service	Sales	DCS
HRC1	HRC2	HRC3	HRC4
Parts Ordered	Overhead Lift	Analyzers	PLC

Supervisor Signature: ______________________________

Date: ______________________________

8 Week Evaluating Supervisor Report

Evaluating Supervisor:____

Date: ____

1. Follows safe work practices, follows rules, does not take short cuts, and understands company safety policies.

    a. Excellent
    b. Very Good
    c. Good
    d. Satisfactory
    e. Unsatisfactory

2. Ability to adapt to and work effectively in a variety of situations and with various individuals and groups. Adapts one's approach as the requirements of the situation change.

    a. Excellent
    b. Very Good
    c. Good
    d. Satisfactory
    e. Unsatisfactory

3. Self-motivated to learn and share ideas. Driven by curiosity and a desire to know more about the unit process, equipment, people and Health, Safety, and Environmental (HSE) initiatives.

    a. Excellent
    b. Very Good
    c. Good
    d. Satisfactory
    e. Unsatisfactory

4. Good attendance. Always at work and on time unless sick or an excusable emergency.

    a. Excellent
    b. Very Good
    c. Good
    d. Satisfactory
    e. Unsatisfactory

5. Gets along with others, wants to be a part of the team as opposed to working sepa4rately. Supports the team concept by information sharing, listening, and contributing ideas.

    a. Excellent
    b. Very Good
    c. Good
    d. Satisfactory
    e. Unsatisfactory

6. Respects the meaning of privileged information. Follows established company procedures.

    a. Excellent
    b. Very Good
    c. Good
    d. Satisfactory
    e. Unsatisfactory

7. Maintains/Prepares satisfactory records. Recognizes tasks that are beyond student capacity. Identifies and attempts to solve discrepancies in systems, results, or information.

    a. Excellent
    b. Very Good
    c. Good
    d. Satisfactory
    e. Unsatisfactory

8. Starts activities in a timely manner. Willingly stays to complete or correct work. Organizes workload.

    a. Excellent
    b. Very Good
    c. Good
    d. Satisfactory
    e. Unsatisfactory

9. Routine tasks are competed within acceptable time. Transfers knowledge of principles and procedures to new techniques. Approaches assignments with confidence.

    a. Excellent
    b. Very Good
    c. Good
    d. Satisfactory
    e. Unsatisfactory

Week 9

Work Order # ______________________________

Work Summary: ______________________________

Duration of Work: ____

Lessons Learned: ____

Work Categories (circle all that apply):

JSA/TSTI/JHA	Hot Work Permit	Excavation Permit	Confined Space Entry
LOTO	Instrumentation	Electrical	Inst. Installation
Elec. Installation	Inst. Removal	Elec. Removal	Inst. Maintenance
Elec. Maintenance	Work Area Cleanup	Inventory Mgt.	Inst. Instruction
Elec. Instruction	Safety Instruction	P&ID	PFD
Wiring Diagram	Motors	DC < 28VDC	DC > 28VDC
AC < 240VAC	AC > 240VAC	Vehicle	Multimeter
Megohmeter	Hand Tools	Power Tools	Tube Bending
Conduit Bending	Conduit Sealing	Communicator	Transmitter calibration
Level Inst.	Pressure Inst.	Temperature Inst.	Flow Inst.
Production	Service	Sales	DCS
HRC1	HRC2	HRC3	HRC4
Parts Ordered	Overhead Lift	Analyzers	PLC

Supervisor Signature: ______________________________

Date: ______________________________

Week 9

Work Order # ______________________________

Work Summary: ______________________________

Duration of Work: ______________________________

Lessons Learned: ______________________________

Work Categories (circle all that apply):

JSA/TSTI/JHA	Hot Work Permit	Excavation Permit	Confined Space Entry
LOTO	Instrumentation	Electrical	Inst. Installation
Elec. Installation	Inst. Removal	Elec. Removal	Inst. Maintenance
Elec. Maintenance	Work Area Cleanup	Inventory Mgt.	Inst. Instruction
Elec. Instruction	Safety Instruction	P&ID	PFD
Wiring Diagram	Motors	DC < 28VDC	DC > 28VDC
AC < 240VAC	AC > 240VAC	Vehicle	Multimeter
Megohmeter	Hand Tools	Power Tools	Tube Bending
Conduit Bending	Conduit Sealing	Communicator	Transmitter calibration
Level Inst.	Pressure Inst.	Temperature Inst.	Flow Inst.
Production	Service	Sales	DCS
HRC1	HRC2	HRC3	HRC4
Parts Ordered	Overhead Lift	Analyzers	PLC

Supervisor Signature: ____

Date: ____

Week 9

Work Order # ____________________

Work Summary: ____________________

Duration of Work: ______________________________

Lessons Learned: ______________________________

Work Categories (circle all that apply):

JSA/TSTI/JHA	Hot Work Permit	Excavation Permit	Confined Space Entry
LOTO	Instrumentation	Electrical	Inst. Installation
Elec. Installation	Inst. Removal	Elec. Removal	Inst. Maintenance
Elec. Maintenance	Work Area Cleanup	Inventory Mgt.	Inst. Instruction
Elec. Instruction	Safety Instruction	P&ID	PFD
Wiring Diagram	Motors	DC < 28VDC	DC > 28VDC
AC < 240VAC	AC > 240VAC	Vehicle	Multimeter
Megohmeter	Hand Tools	Power Tools	Tube Bending
Conduit Bending	Conduit Sealing	Communicator	Transmitter calibration
Level Inst.	Pressure Inst.	Temperature Inst.	Flow Inst.
Production	Service	Sales	DCS
HRC1	HRC2	HRC3	HRC4
Parts Ordered	Overhead Lift	Analyzers	PLC

Supervisor Signature: ______________________________

Date: ______________________________

Week 9

Work Order #

Work Summary:

Duration of Work: ______________________________

Lessons Learned: ______________________________

Work Categories (circle all that apply):

JSA/TSTI/JHA	Hot Work Permit	Excavation Permit	Confined Space Entry
LOTO	Instrumentation	Electrical	Inst. Installation
Elec. Installation	Inst. Removal	Elec. Removal	Inst. Maintenance
Elec. Maintenance	Work Area Cleanup	Inventory Mgt.	Inst. Instruction
Elec. Instruction	Safety Instruction	P&ID	PFD
Wiring Diagram	Motors	DC < 28VDC	DC > 28VDC
AC < 240VAC	AC > 240VAC	Vehicle	Multimeter
Megohmeter	Hand Tools	Power Tools	Tube Bending
Conduit Bending	Conduit Sealing	Communicator	Transmitter calibration
Level Inst.	Pressure Inst.	Temperature Inst.	Flow Inst.
Production	Service	Sales	DCS
HRC1	HRC2	HRC3	HRC4
Parts Ordered	Overhead Lift	Analyzers	PLC

Supervisor Signature: ______________________________

Date: ______________________________

Week 10

Work Order # ______________________

Work Summary: ______________________

Duration of Work: ________________________________

Lessons Learned: ________________________________

Work Categories (circle all that apply):

JSA/TSTI/JHA	Hot Work Permit	Excavation Permit	Confined Space Entry
LOTO	Instrumentation	Electrical	Inst. Installation
Elec. Installation	Inst. Removal	Elec. Removal	Inst. Maintenance
Elec. Maintenance	Work Area Cleanup	Inventory Mgt.	Inst. Instruction
Elec. Instruction	Safety Instruction	P&ID	PFD
Wiring Diagram	Motors	DC < 28VDC	DC > 28VDC
AC < 240VAC	AC > 240VAC	Vehicle	Multimeter
Megohmeter	Hand Tools	Power Tools	Tube Bending
Conduit Bending	Conduit Sealing	Communicator	Transmitter calibration
Level Inst.	Pressure Inst.	Temperature Inst.	Flow Inst.
Production	Service	Sales	DCS
HRC1	HRC2	HRC3	HRC4
Parts Ordered	Overhead Lift	Analyzers	PLC

Supervisor Signature: ______________________________

Date: ______________________________

Week 10

Work Order # ______________________

Work Summary: ______________________

Duration of Work: ____________________

Lessons Learned: ____________________

Work Categories (circle all that apply):

JSA/TSTI/JHA	Hot Work Permit	Excavation Permit	Confined Space Entry
LOTO	Instrumentation	Electrical	Inst. Installation
Elec. Installation	Inst. Removal	Elec. Removal	Inst. Maintenance
Elec. Maintenance	Work Area Cleanup	Inventory Mgt.	Inst. Instruction
Elec. Instruction	Safety Instruction	P&ID	PFD
Wiring Diagram	Motors	DC < 28VDC	DC > 28VDC
AC < 240VAC	AC > 240VAC	Vehicle	Multimeter
Megohmeter	Hand Tools	Power Tools	Tube Bending
Conduit Bending	Conduit Sealing	Communicator	Transmitter calibration
Level Inst.	Pressure Inst.	Temperature Inst.	Flow Inst.
Production	Service	Sales	DCS
HRC1	HRC2	HRC3	HRC4
Parts Ordered	Overhead Lift	Analyzers	PLC

Supervisor Signature: ____

Date: ____

Week 10

Work Order # ____________________

Work Summary: ____________________

Duration of Work: ____________________

Lessons Learned:

Work Categories (circle all that apply):

JSA/TSTI/JHA	Hot Work Permit	Excavation Permit	Confined Space Entry
LOTO	Instrumentation	Electrical	Inst. Installation
Elec. Installation	Inst. Removal	Elec. Removal	Inst. Maintenance
Elec. Maintenance	Work Area Cleanup	Inventory Mgt.	Inst. Instruction
Elec. Instruction	Safety Instruction	P&ID	PFD
Wiring Diagram	Motors	DC < 28VDC	DC > 28VDC
AC < 240VAC	AC > 240VAC	Vehicle	Multimeter
Megohmeter	Hand Tools	Power Tools	Tube Bending
Conduit Bending	Conduit Sealing	Communicator	Transmitter calibration
Level Inst.	Pressure Inst.	Temperature Inst.	Flow Inst.
Production	Service	Sales	DCS
HRC1	HRC2	HRC3	HRC4
Parts Ordered	Overhead Lift	Analyzers	PLC

Supervisor Signature: ______________________________

Date: ______________________________

Week 10

Work Order # ________________

Work Summary: ________________

Duration of Work: ______________________________

Lessons Learned: ______________________________

Work Categories (circle all that apply):

JSA/TSTI/JHA	Hot Work Permit	Excavation Permit	Confined Space Entry
LOTO	Instrumentation	Electrical	Inst. Installation
Elec. Installation	Inst. Removal	Elec. Removal	Inst. Maintenance
Elec. Maintenance	Work Area Cleanup	Inventory Mgt.	Inst. Instruction
Elec. Instruction	Safety Instruction	P&ID	PFD
Wiring Diagram	Motors	DC < 28VDC	DC > 28VDC
AC < 240VAC	AC > 240VAC	Vehicle	Multimeter
Megohmeter	Hand Tools	Power Tools	Tube Bending
Conduit Bending	Conduit Sealing	Communicator	Transmitter calibration
Level Inst.	Pressure Inst.	Temperature Inst.	Flow Inst.
Production	Service	Sales	DCS
HRC1	HRC2	HRC3	HRC4
Parts Ordered	Overhead Lift	Analyzers	PLC

Supervisor Signature: ____

Date: ____

Week 11

Work Order # ____________________

Work Summary: ____________________

Duration of Work: ______________________________

Lessons Learned: ______________________________

Work Categories (circle all that apply):

JSA/TSTI/JHA	Hot Work Permit	Excavation Permit	Confined Space Entry
LOTO	Instrumentation	Electrical	Inst. Installation
Elec. Installation	Inst. Removal	Elec. Removal	Inst. Maintenance
Elec. Maintenance	Work Area Cleanup	Inventory Mgt.	Inst. Instruction
Elec. Instruction	Safety Instruction	P&ID	PFD
Wiring Diagram	Motors	DC < 28VDC	DC > 28VDC
AC < 240VAC	AC > 240VAC	Vehicle	Multimeter
Megohmeter	Hand Tools	Power Tools	Tube Bending
Conduit Bending	Conduit Sealing	Communicator	Transmitter calibration
Level Inst.	Pressure Inst.	Temperature Inst.	Flow Inst.
Production	Service	Sales	DCS
HRC1	HRC2	HRC3	HRC4
Parts Ordered	Overhead Lift	Analyzers	PLC

Supervisor Signature: ______________________________

Date: ______________________________

Week 11

Work Order # ____

Work Summary: ____

Duration of Work: ____

Lessons Learned: ____

Work Categories (circle all that apply):

JSA/TSTI/JHA	Hot Work Permit	Excavation Permit	Confined Space Entry
LOTO	Instrumentation	Electrical	Inst. Installation
Elec. Installation	Inst. Removal	Elec. Removal	Inst. Maintenance
Elec. Maintenance	Work Area Cleanup	Inventory Mgt.	Inst. Instruction
Elec. Instruction	Safety Instruction	P&ID	PFD
Wiring Diagram	Motors	DC < 28VDC	DC > 28VDC
AC < 240VAC	AC > 240VAC	Vehicle	Multimeter
Megohmeter	Hand Tools	Power Tools	Tube Bending
Conduit Bending	Conduit Sealing	Communicator	Transmitter calibration
Level Inst.	Pressure Inst.	Temperature Inst.	Flow Inst.
Production	Service	Sales	DCS
HRC1	HRC2	HRC3	HRC4
Parts Ordered	Overhead Lift	Analyzers	PLC

Supervisor Signature: ______________________________

Date: ______________________________

Week 11

Work Order # ____________________

Work Summary: ____________________

Duration of Work: ______________________________

Lessons Learned: ______________________________

Work Categories (circle all that apply):

JSA/TSTI/JHA	Hot Work Permit	Excavation Permit	Confined Space Entry
LOTO	Instrumentation	Electrical	Inst. Installation
Elec. Installation	Inst. Removal	Elec. Removal	Inst. Maintenance
Elec. Maintenance	Work Area Cleanup	Inventory Mgt.	Inst. Instruction
Elec. Instruction	Safety Instruction	P&ID	PFD
Wiring Diagram	Motors	DC < 28VDC	DC > 28VDC
AC < 240VAC	AC > 240VAC	Vehicle	Multimeter
Megohmeter	Hand Tools	Power Tools	Tube Bending
Conduit Bending	Conduit Sealing	Communicator	Transmitter calibration
Level Inst.	Pressure Inst.	Temperature Inst.	Flow Inst.
Production	Service	Sales	DCS
HRC1	HRC2	HRC3	HRC4
Parts Ordered	Overhead Lift	Analyzers	PLC

Supervisor Signature: ____

Date: ____

Week 11

Work Order # ____________________

Work Summary: ____________________

Duration of Work: ____________________

Lessons Learned: ____________________

Work Categories (circle all that apply):

JSA/TSTI/JHA	Hot Work Permit	Excavation Permit	Confined Space Entry
LOTO	Instrumentation	Electrical	Inst. Installation
Elec. Installation	Inst. Removal	Elec. Removal	Inst. Maintenance
Elec. Maintenance	Work Area Cleanup	Inventory Mgt.	Inst. Instruction
Elec. Instruction	Safety Instruction	P&ID	PFD
Wiring Diagram	Motors	DC < 28VDC	DC > 28VDC
AC < 240VAC	AC > 240VAC	Vehicle	Multimeter
Megohmeter	Hand Tools	Power Tools	Tube Bending
Conduit Bending	Conduit Sealing	Communicator	Transmitter calibration
Level Inst.	Pressure Inst.	Temperature Inst.	Flow Inst.
Production	Service	Sales	DCS
HRC1	HRC2	HRC3	HRC4
Parts Ordered	Overhead Lift	Analyzers	PLC

Supervisor Signature: ______________________________

Date: ______________________________

Week 12

Work Order # ______________________________

Work Summary: ______________________________

Duration of Work: ______________________________

Lessons Learned: ______________________________

Work Categories (circle all that apply):

JSA/TSTI/JHA	Hot Work Permit	Excavation Permit	Confined Space Entry
LOTO	Instrumentation	Electrical	Inst. Installation
Elec. Installation	Inst. Removal	Elec. Removal	Inst. Maintenance
Elec. Maintenance	Work Area Cleanup	Inventory Mgt.	Inst. Instruction
Elec. Instruction	Safety Instruction	P&ID	PFD
Wiring Diagram	Motors	DC < 28VDC	DC > 28VDC
AC < 240VAC	AC > 240VAC	Vehicle	Multimeter
Megohmeter	Hand Tools	Power Tools	Tube Bending
Conduit Bending	Conduit Sealing	Communicator	Transmitter calibration
Level Inst.	Pressure Inst.	Temperature Inst.	Flow Inst.
Production	Service	Sales	DCS
HRC1	HRC2	HRC3	HRC4
Parts Ordered	Overhead Lift	Analyzers	PLC

Supervisor Signature: ______________________________

Date: ______________________________

Week 12

Work Order # ____________________

Work Summary: ____________________

Duration of Work: ______________________________

Lessons Learned: ______________________________

Work Categories (circle all that apply):

JSA/TSTI/JHA	Hot Work Permit	Excavation Permit	Confined Space Entry
LOTO	Instrumentation	Electrical	Inst. Installation
Elec. Installation	Inst. Removal	Elec. Removal	Inst. Maintenance
Elec. Maintenance	Work Area Cleanup	Inventory Mgt.	Inst. Instruction
Elec. Instruction	Safety Instruction	P&ID	PFD
Wiring Diagram	Motors	DC < 28VDC	DC > 28VDC
AC < 240VAC	AC > 240VAC	Vehicle	Multimeter
Megohmeter	Hand Tools	Power Tools	Tube Bending
Conduit Bending	Conduit Sealing	Communicator	Transmitter calibration
Level Inst.	Pressure Inst.	Temperature Inst.	Flow Inst.
Production	Service	Sales	DCS
HRC1	HRC2	HRC3	HRC4
Parts Ordered	Overhead Lift	Analyzers	PLC

Supervisor Signature: ______________________________

Date: ______________________________

Week 12

Work Order # ______________________________

Work Summary: ______________________________

Duration of Work: ____________________

Lessons Learned: ____________________

Work Categories (circle all that apply):

JSA/TSTI/JHA	Hot Work Permit	Excavation Permit	Confined Space Entry
LOTO	Instrumentation	Electrical	Inst. Installation
Elec. Installation	Inst. Removal	Elec. Removal	Inst. Maintenance
Elec. Maintenance	Work Area Cleanup	Inventory Mgt.	Inst. Instruction
Elec. Instruction	Safety Instruction	P&ID	PFD
Wiring Diagram	Motors	DC < 28VDC	DC > 28VDC
AC < 240VAC	AC > 240VAC	Vehicle	Multimeter
Megohmeter	Hand Tools	Power Tools	Tube Bending
Conduit Bending	Conduit Sealing	Communicator	Transmitter calibration
Level Inst.	Pressure Inst.	Temperature Inst.	Flow Inst.
Production	Service	Sales	DCS
HRC1	HRC2	HRC3	HRC4
Parts Ordered	Overhead Lift	Analyzers	PLC

Supervisor Signature: ______________________________

Date: ______________________________

Week 12

Work Order # ______________________________

Work Summary: ______________________________

Duration of Work: ______________________________

Lessons Learned: ______________________________

Work Categories (circle all that apply):

JSA/TSTI/JHA	Hot Work Permit	Excavation Permit	Confined Space Entry
LOTO	Instrumentation	Electrical	Inst. Installation
Elec. Installation	Inst. Removal	Elec. Removal	Inst. Maintenance
Elec. Maintenance	Work Area Cleanup	Inventory Mgt.	Inst. Instruction
Elec. Instruction	Safety Instruction	P&ID	PFD
Wiring Diagram	Motors	DC < 28VDC	DC > 28VDC
AC < 240VAC	AC > 240VAC	Vehicle	Multimeter
Megohmeter	Hand Tools	Power Tools	Tube Bending
Conduit Bending	Conduit Sealing	Communicator	Transmitter calibration
Level Inst.	Pressure Inst.	Temperature Inst.	Flow Inst.
Production	Service	Sales	DCS
HRC1	HRC2	HRC3	HRC4
Parts Ordered	Overhead Lift	Analyzers	PLC

Supervisor Signature: ______________________________

Date: ______________________________

12 Week Evaluating Supervisor Report

Evaluating Supervisor:____

Date: ____

1. Follows safe work practices, follows rules, does not take short cuts, and understands company safety policies.

    a. Excellent
    b. Very Good
    c. Good
    d. Satisfactory
    e. Unsatisfactory

2. Ability to adapt to and work effectively in a variety of situations and with various individuals and groups. Adapts one's approach as the requirements of the situation change.

    a. Excellent
    b. Very Good
    c. Good
    d. Satisfactory
    e. Unsatisfactory

3. Self-motivated to learn and share ideas. Driven by curiosity and a desire to know more about the unit process, equipment, people and Health, Safety, and Environmental (HSE) initiatives.

    a. Excellent
    b. Very Good
    c. Good
    d. Satisfactory
    e. Unsatisfactory

4. Good attendance. Always at work and on time unless sick or an excusable emergency.

    a. Excellent
    b. Very Good
    c. Good
    d. Satisfactory
    e. Unsatisfactory

5. Gets along with others, wants to be a part of the team as opposed to working sepa4rately. Supports the team concept by information sharing, listening, and contributing ideas.

    a. Excellent
    b. Very Good
    c. Good
    d. Satisfactory
    e. Unsatisfactory

6. Respects the meaning of privileged information. Follows established company procedures.

    a. Excellent
    b. Very Good
    c. Good
    d. Satisfactory
    e. Unsatisfactory

7. Maintains/Prepares satisfactory records. Recognizes tasks that are beyond student capacity. Identifies and attempts to solve discrepancies in systems, results, or information.

    a. Excellent
    b. Very Good
    c. Good
    d. Satisfactory
    e. Unsatisfactory

8. Starts activities in a timely manner. Willingly stays to complete or correct work. Organizes workload.

    a. Excellent
    b. Very Good
    c. Good
    d. Satisfactory
    e. Unsatisfactory

9. Routine tasks are competed within acceptable time. Transfers knowledge of principles and procedures to new techniques. Approaches assignments with confidence.

    a. Excellent
    b. Very Good
    c. Good
    d. Satisfactory
    e. Unsatisfactory

Week 13

Work Order # ______________________________

Work Summary: ______________________________

Duration of Work:

Lessons Learned:

Work Categories (circle all that apply):

JSA/TSTI/JHA	Hot Work Permit	Excavation Permit	Confined Space Entry
LOTO	Instrumentation	Electrical	Inst. Installation
Elec. Installation	Inst. Removal	Elec. Removal	Inst. Maintenance
Elec. Maintenance	Work Area Cleanup	Inventory Mgt.	Inst. Instruction
Elec. Instruction	Safety Instruction	P&ID	PFD
Wiring Diagram	Motors	DC < 28VDC	DC > 28VDC
AC < 240VAC	AC > 240VAC	Vehicle	Multimeter
Megohmeter	Hand Tools	Power Tools	Tube Bending
Conduit Bending	Conduit Sealing	Communicator	Transmitter calibration
Level Inst.	Pressure Inst.	Temperature Inst.	Flow Inst.
Production	Service	Sales	DCS
HRC1	HRC2	HRC3	HRC4
Parts Ordered	Overhead Lift	Analyzers	PLC

Supervisor Signature: ______________________________

Date: ______________________________

Week 13

Work Order #

Work Summary:

Duration of Work: ______________________________

Lessons Learned: ______________________________

Work Categories (circle all that apply):

JSA/TSTI/JHA	Hot Work Permit	Excavation Permit	Confined Space Entry
LOTO	Instrumentation	Electrical	Inst. Installation
Elec. Installation	Inst. Removal	Elec. Removal	Inst. Maintenance
Elec. Maintenance	Work Area Cleanup	Inventory Mgt.	Inst. Instruction
Elec. Instruction	Safety Instruction	P&ID	PFD
Wiring Diagram	Motors	DC < 28VDC	DC > 28VDC
AC < 240VAC	AC > 240VAC	Vehicle	Multimeter
Megohmeter	Hand Tools	Power Tools	Tube Bending
Conduit Bending	Conduit Sealing	Communicator	Transmitter calibration
Level Inst.	Pressure Inst.	Temperature Inst.	Flow Inst.
Production	Service	Sales	DCS
HRC1	HRC2	HRC3	HRC4
Parts Ordered	Overhead Lift	Analyzers	PLC

Supervisor Signature: ____

Date: ____

Week 13

Work Order # ____

Work Summary: ____

Duration of Work: ____

Lessons Learned: ____

Work Categories (circle all that apply):

JSA/TSTI/JHA	Hot Work Permit	Excavation Permit	Confined Space Entry
LOTO	Instrumentation	Electrical	Inst. Installation
Elec. Installation	Inst. Removal	Elec. Removal	Inst. Maintenance
Elec. Maintenance	Work Area Cleanup	Inventory Mgt.	Inst. Instruction
Elec. Instruction	Safety Instruction	P&ID	PFD
Wiring Diagram	Motors	DC < 28VDC	DC > 28VDC
AC < 240VAC	AC > 240VAC	Vehicle	Multimeter
Megohmeter	Hand Tools	Power Tools	Tube Bending
Conduit Bending	Conduit Sealing	Communicator	Transmitter calibration
Level Inst.	Pressure Inst.	Temperature Inst.	Flow Inst.
Production	Service	Sales	DCS
HRC1	HRC2	HRC3	HRC4
Parts Ordered	Overhead Lift	Analyzers	PLC

Supervisor Signature: ______________________________

Date: ______________________________

Week 13

Work Order # ______________________________

Work Summary: ______________________________

Duration of Work: ______________________________

Lessons Learned: ______________________________

Work Categories (circle all that apply):

JSA/TSTI/JHA	Hot Work Permit	Excavation Permit	Confined Space Entry
LOTO	Instrumentation	Electrical	Inst. Installation
Elec. Installation	Inst. Removal	Elec. Removal	Inst. Maintenance
Elec. Maintenance	Work Area Cleanup	Inventory Mgt.	Inst. Instruction
Elec. Instruction	Safety Instruction	P&ID	PFD
Wiring Diagram	Motors	DC < 28VDC	DC > 28VDC
AC < 240VAC	AC > 240VAC	Vehicle	Multimeter
Megohmeter	Hand Tools	Power Tools	Tube Bending
Conduit Bending	Conduit Sealing	Communicator	Transmitter calibration
Level Inst.	Pressure Inst.	Temperature Inst.	Flow Inst.
Production	Service	Sales	DCS
HRC1	HRC2	HRC3	HRC4
Parts Ordered	Overhead Lift	Analyzers	PLC

Supervisor Signature: ______________________________

Date: ______________________________

Week 14

Work Order # ______________________________

Work Summary: ______________________________

Duration of Work: ______________________________

Lessons Learned: ______________________________

Work Categories (circle all that apply):

JSA/TSTI/JHA	Hot Work Permit	Excavation Permit	Confined Space Entry
LOTO	Instrumentation	Electrical	Inst. Installation
Elec. Installation	Inst. Removal	Elec. Removal	Inst. Maintenance
Elec. Maintenance	Work Area Cleanup	Inventory Mgt.	Inst. Instruction
Elec. Instruction	Safety Instruction	P&ID	PFD
Wiring Diagram	Motors	DC < 28VDC	DC > 28VDC
AC < 240VAC	AC > 240VAC	Vehicle	Multimeter
Megohmeter	Hand Tools	Power Tools	Tube Bending
Conduit Bending	Conduit Sealing	Communicator	Transmitter calibration
Level Inst.	Pressure Inst.	Temperature Inst.	Flow Inst.
Production	Service	Sales	DCS
HRC1	HRC2	HRC3	HRC4
Parts Ordered	Overhead Lift	Analyzers	PLC

Supervisor Signature: ______________________________

Date: ______________________________

Week 14

Work Order # ________________________

Work Summary: ________________________

Duration of Work: ______________________________

Lessons Learned: ______________________________

Work Categories (circle all that apply):

JSA/TSTI/JHA	Hot Work Permit	Excavation Permit	Confined Space Entry
LOTO	Instrumentation	Electrical	Inst. Installation
Elec. Installation	Inst. Removal	Elec. Removal	Inst. Maintenance
Elec. Maintenance	Work Area Cleanup	Inventory Mgt.	Inst. Instruction
Elec. Instruction	Safety Instruction	P&ID	PFD
Wiring Diagram	Motors	DC < 28VDC	DC > 28VDC
AC < 240VAC	AC > 240VAC	Vehicle	Multimeter
Megohmeter	Hand Tools	Power Tools	Tube Bending
Conduit Bending	Conduit Sealing	Communicator	Transmitter calibration
Level Inst.	Pressure Inst.	Temperature Inst.	Flow Inst.
Production	Service	Sales	DCS
HRC1	HRC2	HRC3	HRC4
Parts Ordered	Overhead Lift	Analyzers	PLC

Supervisor Signature: ______________________________

Date: ______________________________

Week 14

Work Order # ____________________

Work Summary:

Duration of Work: ______________________________

Lessons Learned: ______________________________

Work Categories (circle all that apply):

JSA/TSTI/JHA	Hot Work Permit	Excavation Permit	Confined Space Entry
LOTO	Instrumentation	Electrical	Inst. Installation
Elec. Installation	Inst. Removal	Elec. Removal	Inst. Maintenance
Elec. Maintenance	Work Area Cleanup	Inventory Mgt.	Inst. Instruction
Elec. Instruction	Safety Instruction	P&ID	PFD
Wiring Diagram	Motors	DC < 28VDC	DC > 28VDC
AC < 240VAC	AC > 240VAC	Vehicle	Multimeter
Megohmeter	Hand Tools	Power Tools	Tube Bending
Conduit Bending	Conduit Sealing	Communicator	Transmitter calibration
Level Inst.	Pressure Inst.	Temperature Inst.	Flow Inst.
Production	Service	Sales	DCS
HRC1	HRC2	HRC3	HRC4
Parts Ordered	Overhead Lift	Analyzers	PLC

Supervisor Signature: ______________________________

Date: ______________________________

Week 14

Work Order #

Work Summary:

Duration of Work: ____________________

Lessons Learned: ____________________

Work Categories (circle all that apply):

JSA/TSTI/JHA	Hot Work Permit	Excavation Permit	Confined Space Entry
LOTO	Instrumentation	Electrical	Inst. Installation
Elec. Installation	Inst. Removal	Elec. Removal	Inst. Maintenance
Elec. Maintenance	Work Area Cleanup	Inventory Mgt.	Inst. Instruction
Elec. Instruction	Safety Instruction	P&ID	PFD
Wiring Diagram	Motors	DC < 28VDC	DC > 28VDC
AC < 240VAC	AC > 240VAC	Vehicle	Multimeter
Megohmeter	Hand Tools	Power Tools	Tube Bending
Conduit Bending	Conduit Sealing	Communicator	Transmitter calibration
Level Inst.	Pressure Inst.	Temperature Inst.	Flow Inst.
Production	Service	Sales	DCS
HRC1	HRC2	HRC3	HRC4
Parts Ordered	Overhead Lift	Analyzers	PLC

Supervisor Signature: ____

Date: ____

Week 15

Work Order # ____________________

Work Summary: ____________________

Duration of Work: ______________________________

Lessons Learned: ______________________________

Work Categories (circle all that apply):

JSA/TSTI/JHA	Hot Work Permit	Excavation Permit	Confined Space Entry
LOTO	Instrumentation	Electrical	Inst. Installation
Elec. Installation	Inst. Removal	Elec. Removal	Inst. Maintenance
Elec. Maintenance	Work Area Cleanup	Inventory Mgt.	Inst. Instruction
Elec. Instruction	Safety Instruction	P&ID	PFD
Wiring Diagram	Motors	DC < 28VDC	DC > 28VDC
AC < 240VAC	AC > 240VAC	Vehicle	Multimeter
Megohmeter	Hand Tools	Power Tools	Tube Bending
Conduit Bending	Conduit Sealing	Communicator	Transmitter calibration
Level Inst.	Pressure Inst.	Temperature Inst.	Flow Inst.
Production	Service	Sales	DCS
HRC1	HRC2	HRC3	HRC4
Parts Ordered	Overhead Lift	Analyzers	PLC

Supervisor Signature: ______________________________

Date: ______________________________

Week 15

Work Order # ________________________________

Work Summary: ________________________________

Duration of Work: ____________________

Lessons Learned: ____________________

Work Categories (circle all that apply):

JSA/TSTI/JHA	Hot Work Permit	Excavation Permit	Confined Space Entry
LOTO	Instrumentation	Electrical	Inst. Installation
Elec. Installation	Inst. Removal	Elec. Removal	Inst. Maintenance
Elec. Maintenance	Work Area Cleanup	Inventory Mgt.	Inst. Instruction
Elec. Instruction	Safety Instruction	P&ID	PFD
Wiring Diagram	Motors	DC < 28VDC	DC > 28VDC
AC < 240VAC	AC > 240VAC	Vehicle	Multimeter
Megohmeter	Hand Tools	Power Tools	Tube Bending
Conduit Bending	Conduit Sealing	Communicator	Transmitter calibration
Level Inst.	Pressure Inst.	Temperature Inst.	Flow Inst.
Production	Service	Sales	DCS
HRC1	HRC2	HRC3	HRC4
Parts Ordered	Overhead Lift	Analyzers	PLC

Supervisor Signature: ____

Date: ____

Week 15

Work Order # ______________________________

Work Summary: ______________________________

Duration of Work: ______________________________

Lessons Learned: ______________________________

Work Categories (circle all that apply):

JSA/TSTI/JHA	Hot Work Permit	Excavation Permit	Confined Space Entry
LOTO	Instrumentation	Electrical	Inst. Installation
Elec. Installation	Inst. Removal	Elec. Removal	Inst. Maintenance
Elec. Maintenance	Work Area Cleanup	Inventory Mgt.	Inst. Instruction
Elec. Instruction	Safety Instruction	P&ID	PFD
Wiring Diagram	Motors	DC < 28VDC	DC > 28VDC
AC < 240VAC	AC > 240VAC	Vehicle	Multimeter
Megohmeter	Hand Tools	Power Tools	Tube Bending
Conduit Bending	Conduit Sealing	Communicator	Transmitter calibration
Level Inst.	Pressure Inst.	Temperature Inst.	Flow Inst.
Production	Service	Sales	DCS
HRC1	HRC2	HRC3	HRC4
Parts Ordered	Overhead Lift	Analyzers	PLC

Supervisor Signature: ______________________________

Date: ______________________________

Week 15

Work Order #

Work Summary:

Duration of Work: ____________________

Lessons Learned: ____________________

Work Categories (circle all that apply):

JSA/TSTI/JHA	Hot Work Permit	Excavation Permit	Confined Space Entry
LOTO	Instrumentation	Electrical	Inst. Installation
Elec. Installation	Inst. Removal	Elec. Removal	Inst. Maintenance
Elec. Maintenance	Work Area Cleanup	Inventory Mgt.	Inst. Instruction
Elec. Instruction	Safety Instruction	P&ID	PFD
Wiring Diagram	Motors	DC < 28VDC	DC > 28VDC
AC < 240VAC	AC > 240VAC	Vehicle	Multimeter
Megohmeter	Hand Tools	Power Tools	Tube Bending
Conduit Bending	Conduit Sealing	Communicator	Transmitter calibration
Level Inst.	Pressure Inst.	Temperature Inst.	Flow Inst.
Production	Service	Sales	DCS
HRC1	HRC2	HRC3	HRC4
Parts Ordered	Overhead Lift	Analyzers	PLC

Supervisor Signature: ______________________________

Date: ______________________________

Week 16

Work Order # ______________________________

Work Summary: ______________________________

Duration of Work: ____________________

Lessons Learned: ____________________

Work Categories (circle all that apply):

JSA/TSTI/JHA	Hot Work Permit	Excavation Permit	Confined Space Entry
LOTO	Instrumentation	Electrical	Inst. Installation
Elec. Installation	Inst. Removal	Elec. Removal	Inst. Maintenance
Elec. Maintenance	Work Area Cleanup	Inventory Mgt.	Inst. Instruction
Elec. Instruction	Safety Instruction	P&ID	PFD
Wiring Diagram	Motors	DC < 28VDC	DC > 28VDC
AC < 240VAC	AC > 240VAC	Vehicle	Multimeter
Megohmeter	Hand Tools	Power Tools	Tube Bending
Conduit Bending	Conduit Sealing	Communicator	Transmitter calibration
Level Inst.	Pressure Inst.	Temperature Inst.	Flow Inst.
Production	Service	Sales	DCS
HRC1	HRC2	HRC3	HRC4
Parts Ordered	Overhead Lift	Analyzers	PLC

Supervisor Signature: ______________________________

Date: ______________________________

Week 16

Work Order # ______________________________

Work Summary: ______________________________

Duration of Work: ______________________________

Lessons Learned: ______________________________

Work Categories (circle all that apply):

JSA/TSTI/JHA	Hot Work Permit	Excavation Permit	Confined Space Entry
LOTO	Instrumentation	Electrical	Inst. Installation
Elec. Installation	Inst. Removal	Elec. Removal	Inst. Maintenance
Elec. Maintenance	Work Area Cleanup	Inventory Mgt.	Inst. Instruction
Elec. Instruction	Safety Instruction	P&ID	PFD
Wiring Diagram	Motors	DC < 28VDC	DC > 28VDC
AC < 240VAC	AC > 240VAC	Vehicle	Multimeter
Megohmeter	Hand Tools	Power Tools	Tube Bending
Conduit Bending	Conduit Sealing	Communicator	Transmitter calibration
Level Inst.	Pressure Inst.	Temperature Inst.	Flow Inst.
Production	Service	Sales	DCS
HRC1	HRC2	HRC3	HRC4
Parts Ordered	Overhead Lift	Analyzers	PLC

Supervisor Signature: ____

Date: ____

Week 16

Work Order # ______________________________

Work Summary: ______________________________

Duration of Work: ______________________________

Lessons Learned: ______________________________

Work Categories (circle all that apply):

JSA/TSTI/JHA	Hot Work Permit	Excavation Permit	Confined Space Entry
LOTO	Instrumentation	Electrical	Inst. Installation
Elec. Installation	Inst. Removal	Elec. Removal	Inst. Maintenance
Elec. Maintenance	Work Area Cleanup	Inventory Mgt.	Inst. Instruction
Elec. Instruction	Safety Instruction	P&ID	PFD
Wiring Diagram	Motors	DC < 28VDC	DC > 28VDC
AC < 240VAC	AC > 240VAC	Vehicle	Multimeter
Megohmeter	Hand Tools	Power Tools	Tube Bending
Conduit Bending	Conduit Sealing	Communicator	Transmitter calibration
Level Inst.	Pressure Inst.	Temperature Inst.	Flow Inst.
Production	Service	Sales	DCS
HRC1	HRC2	HRC3	HRC4
Parts Ordered	Overhead Lift	Analyzers	PLC

Supervisor Signature: ______________________________

Date: ______________________________

Week 16

Work Order # ____

Work Summary: ____

Duration of Work: ____________________

Lessons Learned: ____________________

Work Categories (circle all that apply):

JSA/TSTI/JHA	Hot Work Permit	Excavation Permit	Confined Space Entry
LOTO	Instrumentation	Electrical	Inst. Installation
Elec. Installation	Inst. Removal	Elec. Removal	Inst. Maintenance
Elec. Maintenance	Work Area Cleanup	Inventory Mgt.	Inst. Instruction
Elec. Instruction	Safety Instruction	P&ID	PFD
Wiring Diagram	Motors	DC < 28VDC	DC > 28VDC
AC < 240VAC	AC > 240VAC	Vehicle	Multimeter
Megohmeter	Hand Tools	Power Tools	Tube Bending
Conduit Bending	Conduit Sealing	Communicator	Transmitter calibration
Level Inst.	Pressure Inst.	Temperature Inst.	Flow Inst.
Production	Service	Sales	DCS
HRC1	HRC2	HRC3	HRC4
Parts Ordered	Overhead Lift	Analyzers	PLC

Supervisor Signature: ____

Date: ____

16 Week Evaluating Supervisor Report

Evaluating Supervisor:______________________________

Date: ______________________________

1. Follows safe work practices, follows rules, does not take short cuts, and understands company safety policies.

    f. Excellent
    g. Very Good
    h. Good
    i. Satisfactory
    j. Unsatisfactory

2. Ability to adapt to and work effectively in a variety of situations and with various individuals and groups. Adapts one's approach as the requirements of the situation change.

    a. Excellent
    b. Very Good
    c. Good
    d. Satisfactory
    e. Unsatisfactory

3. Self-motivated to learn and share ideas. Driven by curiosity and a desire to know more about the unit process, equipment, people and Health, Safety, and Environmental (HSE) initiatives.

    a. Excellent
    b. Very Good
    c. Good
    d. Satisfactory
    e. Unsatisfactory

4. Good attendance. Always at work and on time unless sick or an excusable emergency.

    a. Excellent
    b. Very Good
    c. Good
    d. Satisfactory
    e. Unsatisfactory

5. Gets along with others, wants to be a part of the team as opposed to working sepa4rately. Supports the team concept by information sharing, listening, and contributing ideas.

    a. Excellent
    b. Very Good
    c. Good
    d. Satisfactory
    e. Unsatisfactory

6. Respects the meaning of privileged information. Follows established company procedures.

    a. Excellent
    b. Very Good
    c. Good
    d. Satisfactory
    e. Unsatisfactory

7. Maintains/Prepares satisfactory records. Recognizes tasks that are beyond student capacity. Identifies and attempts to solve discrepancies in systems, results, or information.

    a. Excellent
    b. Very Good
    c. Good
    d. Satisfactory
    e. Unsatisfactory

8. Starts activities in a timely manner. Willingly stays to complete or correct work. Organizes workload.

    a. Excellent
    b. Very Good
    c. Good
    d. Satisfactory
    e. Unsatisfactory

9. Routine tasks are competed within acceptable time. Transfers knowledge of principles and procedures to new techniques. Approaches assignments with confidence.

    a. Excellent
    b. Very Good
    c. Good
    d. Satisfactory
    e. Unsatisfactory

Made in the USA
Coppell, TX
05 February 2020